Lost Airline Liveries

AIRLINE COLOR SCHEMES OF THE PAST

Lost Airline Liveries

JOHN K. MORTON

Motorbooks International
Publishers & Wholesalers

INTRODUCTION

Lost Airlines – Airline Colour Schemes of the Past, is a companion volume to my highly successful book *Faded Glory* published by Airlife in 1991. The interest shown in the book prompted me to put together another collection of my pictures and present them along with a brief history of the carrier, enabling the reader to keep a record of airlines and colours that were once to be seen at airports around the world.

The majority of the airlines included in this volume have disappeared completely, whilst those remaining have adopted a new image and colours.

All the illustrations in *Lost Airlines* were taken by myself using Kodachrome 25 film, and I am obviously very much indebted to the management of many of the world's airports for their help that enabled me to obtain the pictures to make this book possible.

John K. Morton

This edition first published in 1996 by Motorbooks International, Publishers & Wholesalers, PO Box 2, 729 Prospect Avenue, Osceola, WI 54020, USA.

© John K. Morton 1996

Previously published by Airlife Publishing Ltd., Shrewsbury, England, 1996

Library of Congress Cataloging-in-Publication Data is available

ISBN 0-7603–0258-8

Printed in Italy

CAL AIR

Formerly known as British Caledonian Airways (Charter) Ltd, the Company changed its name towards the end of 1985 to become Cal Air. At this time a new livery was introduced, shown here on a Series 10 McDonnell Douglas DC-10 arriving at London's Gatwick Airport in August 1987. The traditional Lion motif previously seen on all British Caledonian aircraft was retained on the Cal Air colour scheme, appearing in red on a white ground instead of the original gold on a blue ground.

NOVAIR

Cal Air changed its name to Novair International Airways Ltd at the beginning of 1989. The basic colour scheme was retained, the main changes being the replacement of the Cal Air title by Novair, and the removal of the Caledonian Lion motif from the tail. McDonnell Douglas DC-10 Series 10 G-BJZE is about to depart Manchester Airport in May 1989, at the start of a charter flight to a Mediterranean destination. The airline's parent company, the Rank Organisation, intended to sell the carrier as a going concern at the beginning of 1990, but was unable to find a buyer, and Novair ceased operations in May of that year. At that time the airline had both Boeing 737s and McDonnell Douglas DC-10s in its fleet.

LONDON CITY AIRWAYS

Founded in 1986 and formerly known as Eurocity Express Ltd, this airline commenced operations in 1987 with a fleet of two De Havilland Dash 7s. Based at London City Airport, it was one of the first carriers to operate from the new STOL terminal. In the spring of 1988 the company's name was changed to London City Airways, and at the same time the carrier inaugurated a new scheduled service from London City to Amsterdam. It merged into British Midland Airways in September 1990. One of the airline's Dash 7s, bearing the title London City Airways, is seen at East Midlands Airport in June 1989.

CAPITAL AIRLINES

Capital Airlines, originally known as Brown Air International, was formed in 1983 and based at Leeds/Bradford Airport in the north of England. It started services with a staff of four and a nine-seater Cessna, later adding a Gulfstream to the fleet before becoming known as Capital Airlines in 1987. At this time 39-seater Shorts 360 aircraft were put into service, and continued flying on scheduled services from Leeds/Bradford to London Gatwick until operations ceased in June 1990, following the demise of parent company Brown Group International. Shorts 360 G-OLBA, photographed at the carrier's base in June 1989, is about to depart on the first flight of the day to London Gatwick.

HOLIDAIR AIRWAYS

Holidair Airways was a Canadian charter airline based in Calgary, Alberta, with a fleet of McDonnell Douglas DC-8s. Its major operations were for a Canadian tour company which collapsed, forcing the airline to cease flying in October 1989 owing to financial difficulties. McDonnell Douglas DC-8 C-FHAB, wearing the full colour scheme of Holidair, is seen arriving at Vancouver in August 1989, a few weeks before the airline ceased operations.

AERONICA

Founded in 1981, Aeronica was wholly owned and controlled by the government of Nicaragua. From its base at Managua in Central America the airline operated scheduled flights, including a daily service to Miami, and Boeing 727 Series 25 YN-BXW is seen arriving at the North American city in December 1990.

In April 1991 Aeronica leased Boeing 707 YN-CDE, which flew for the airline for about twelve months. In December 1991 it was a regular visitor to Florida, making almost daily arrivals and departures, and is seen here about to touch down at Miami Airport. At the end of the leasing period the aircraft was withdrawn from service. In the early part of 1992 Aeronica was purchased by the parent company of TACA, the international airline of the Republic of El Salvador.

UTA

UTA (Union de Transports Aériens) was formed in Paris in October 1963, and in its later years operated long-haul services from France to Africa, the Far East, Australia and New Zealand with a modern fleet of McDonnell Douglas DC-10s and Boeing 747s. On 1 January 1992 UTA's merger into Air France was announced, and this was completed by December of the same year. Boeing 747 Series 400 F-GEXA was photographed at Sydney airport whilst still flying in the UTA colour scheme. It is being pushed back from the gate at the start of its flight to Paris in April 1990.

TRUMP AIRLINES

It is 05.45 in May 1990 as the sun rises over the East River, adjacent to New York's La Guardia Airport, and the Trump Shuttle terminal is being prepared for the day's flights. Trump Airlines, founded by Donald Trump, began operations in June 1989 and was based at La Guardia. The airline took over the existing services provided by Eastern Airlines on shuttle flights between New York and Boston, the Eastern titles and colours being replaced by those of Trump. Trump Airlines ran into financial difficulties and entered into an agreement with USAir to take over the operation of Trump Shuttle, and this became effective in April 1992.

Boeing 727s, previously flown by Eastern Airlines and now in the colours of Trump, stand at their gates at La Guardia Airport in May 1990. The aircraft are in all-economy class configuration.

PAN AMERICAN WORLD AIRWAYS

Pan American World Airways must surely have been one of the most well-known and famous airlines. No visit to any major airport around the world could have failed to produce the sight of at least one of its aircraft. Pan Am was one of the world's major carriers, with a route network taking its colours around the globe. Founded in 1927 by Juan Trippe, the airline was the first to introduce an around-the-world service, which began in 1947, using Lockheed Constellations. The route followed an easterly direction and took fourteen days to complete the New York-New York round trip. As more and more people travelled, Pan Am recognised the need for larger aircraft, and the airline was instrumental in the introduction of the Boeing 747. Indeed, it was at Pan Am's insistence that Boeing design the now well-known aircraft. Like many other airlines, Pan Am suffered financial difficulties, serious enough to oblige it to sell its Pan Am skyscraper building in Manhattan. However, the company continued to lose money, and in the summer of 1991 a deal was made with Delta Air Lines, the Atlanta based carrier, to take over the airline and most of its routes. As Delta was not in any position to absorb the continuing losses, Pan Am ceased operations completely on 4 December 1991.

McDonnell Douglas DC-10 N61NA, a Series 10 model joined Pan Am's fleet after the purchase of Florida-based National Airlines. The trijet is seen on finals to Miami Airport in October 1982.

Boeing 747 Series 100 N4703U, photographed at Miami in December 1988, came into the Pan Am fleet in January 1986, having previously flown in the colours of United Airlines.

This Series 100 Boeing 747, N737PA, built by Boeing in January 1970, was photographed as it was about to turn on to the San Francisco runway in August 1990.

Pan Am had its own terminal at Kennedy Airport, New York, and this picture, taken in May 1990, shows a collection of Airbus Industrie A310s being prepared for departure at their respective stands.

Pan Am used Airbus A310s on intercontinental services out of Florida. Here N208PA, leased from Airbus, catches the last rays of the sun as it turns on to runway 9L at Miami International in December 1990.

Pan Am operated shuttle services to Washington and Boston from New York, these flights being operated by Boeing 727s. Having dispensed with the services of the push tractor, N375PA is about to taxi to the runway at JFK Airport for an evening flight in October 1990.

Premium-fare-paying passengers were offered a helicopter fast connection service from Kennedy Airport to downtown New York, but these flights were discontinued following the fatal accident of a helicopter soon after landing on the roof of the Pan Am building in mid-town Manhattan. Bell 222 N3899G is seen about to depart JFK in June 1983.

SUNWORLD INTERNATIONAL

A Las Vegas based airline, Sunworld International Airways commenced services in May 1983, operating a fleet of McDonnell Douglas DC-9s to destinations in the western states of the USA. Boeing 737s were later added to the fleet for a short period, but these were returned and the carrier continued to provide services with its original DC-9s. Sunworld entered Chapter 11 bankruptcy protection, and finally ceased operations in November 1988. In this picture DC-9 Series 14 N9102 is seen parked at McCarron International Airport, its home base, in July 1988.

SKYWORLD

Originally known as Ports of Call Air, this Denver, Colorado, based travel club began operations in 1968 and provided extensive internal and overseas flights for its vast membership, together with charter operations. Its original fleet of Convair 990s was later replaced by Boeing 707s, all of which were configured in economy class. Titles carried by the airline's fleet showed 'Ports of Call Denver' on the fuselage, but these were replaced with 'Skyworld' at the end of 1986, when the travel club became known by this name. Boeing 707 N712PC was operating a flight out of Las Vegas when photographed in August 1988. The airline continued flying for a further year, ceasing operations in July 1989.

PRESIDENTIAL EXPRESS

Presidential Express Jetstream 31 N104XV, photographed at Washington Dulles Airport in October 1988, was operated on behalf of United Airlines to provide feeder services. The Virginia based airline was a subsidiary of Washington based Presidential Airways Inc, and had a fleet of ten of these aircraft. The company filed for Chapter 11 bankruptcy protection in October 1989, and ceased operations completely in December of that year.

PACIFIC INTERSTATE AIRLINES

Boeing 727 N5609, bearing Bahamas Express titles, was in fact operated by the Las Vegas based company Pacific Interstate Airlines Inc. The Boeing was used on charter flights to the Bahamas, and was photographed arriving at Atlanta, Georgia, in April 1988. Operations ceased soon after this picture was taken, the aircraft and services being sold to Fort Lauderdale, Florida, based Majestic Air.

MAJESTIC AIR

After the sale had taken place, N5609 appeared in the livery of Majestic Air, and is seen here about to depart Fort Lauderdale for the Bahamas in December 1988. The airline continued operating under this name for a further few months, later becoming known as Carnival Airlines. This aircraft continued to fly with Carnival until it was withdrawn from service in the summer of 1991.

GULF AIR TRANSPORT

Gulf Air Transport specialised in passenger charter services in the Americas. Founded in 1979 and based in Louisiana, the company later became known as Gulf Air Inc, and operated a fleet of Boeing 727s and, later, McDonnell Douglas DC-8s. In July 1989 the company was renamed Transocean Airways Inc, and continued in business under this title until it ceased operations in the first few months of 1990, after being forced into bankruptcy. Boeing 727 N504AV is seen arriving at Miami in October 1988.

ODYSSEY INTERNATIONAL

Odyssey International was originally established in Ontario, Canada, in 1988 as a charter airline, and services commenced during the winter of 1988/89, using a Boeing 757 leased from Air Holland. The aircraft remained in the colours of the Dutch carrier, but with Odyssey titles added, as illustrated in this study of PH-AHF about to return to Canada from Fort Lauderdale in December 1988.

The airline continued to take delivery of Boeing 757s in the early part of 1989, also adding 737s to its fleet. As new aircraft were brought into service the Odyssey colours were introduced, and are shown here on Boeing 757 C-GAWB, arriving at Manchester from Toronto in April 1990. A few days after this photograph was taken the parent company went into receivership and the airline ceased operating.

AIR EUROPE

Founded in 1978 and based at London Gatwick, Air Europe was one of the United Kingdom's major holiday airlines, providing flights for the principal UK tour operators. Services started with a small fleet of Boeing 737 Series 200s, and such was the success of the airline that it was found necessary to enlarge the fleet and introduce the new 300 and 400 Series 737s and the Boeing 757. The airline then entered the scheduled flight business to complement their charter services. An impressive line-up of Air Europe 737s were about to depart for various European destinations when captured by the camera at London Gatwick in June 1989.

Air Europe accepted delivery of its Boeing 757s in March 1983, and they were immediately put into service on routes to popular destinations in Spain and Greece. The first of the type to be delivered G-BKRM was photographed in August 1983, about to touch down on the island of Corfu.

By 1987 Air Europe had considerably enlarged its fleet, and its colours were to be seen at most of the British airports from where holidaymakers departed for the sunspots of Europe. Boeing 737 Series 300 G-BMTF was returning to the UK when photographed on final approach to Gatwick Airport in August of that year.

In the summer of 1989 the carrier found it necessary to lease a Boeing 747 to provide additional capacity during the busy holiday season. A Series 100 model was leased from the American airline Tower Air, and, while it did not appear in the full colours of Air Europe, it did receive titles. The 1970-vintage airliner retained its American registration, N602FF, while in service on this side of the Atlantic, and is seen here about to approach its stand at Manchester in June 1989.

DAN-AIR

Owned by the shipping company Davies & Newman, from which it derived its name, Dan-Air was one of the United Kingdom's inclusive-tour charter airlines which operated out of many of Britain's airports to various European destinations. The airline began operations in 1953, and had a varied selection of aircraft in its fleet. The liveries of the carrier's aircraft underwent minor changes over the years, as will be observed in the illustrations in this spread. Boeing 727 Series 46 G-BAFZ, built in 1966, was originally delivered to and flown by Japan Airlines. The early Dan-Air red and black livery was later changed to red and blue, retaining the flag within a circle on the tail. The 727 was photographed while being pushed back from the gate at Manchester Airport in June 1978.

Like many other airlines, Dan-Air leased its aircraft to other carriers during quiet periods, and Boeing 727 G-BHVT, photographed at Manchester on a very wet afternoon in June 1983, wears the red and blue colours of Lacsa, the central American airline based in Costa Rica. For a period of four years from 1980 this aircraft flew for Dan-Air during the summer months and for Lacsa in the winter.

In 1984 the airline leased-in several Boeing 737s to provide additional capacity during the United Kingdom's busy holiday period. One of the aircraft on lease G-BJXJ, was photographed as it was about to turn and take off from Corfu airport in August 1983, and illustrates the colours which Dan-Air used for a further five years.

This is not a regular Dan-Air colour scheme, but depicts Airbus A300 G-BMNA in the livery of the German airline Hapag Lloyd. It was leased to Dan-Air in the summer of 1986, and is seen taking off from Palma in May of that year.

One of the carrier's winter leases involved the transfer of Boeing 727 G-BHVT to the North American airline Sun Country. The aircraft has now lost its Lacsa-style colours, and carries the regular Dan-Air scheme, although the Dan-Air titles have been replaced by those of Sun Country. This picture was taken at Las Vegas in April 1985. The 727 returned to the UK each summer until 1990, spending each winter flying for Sun Country.

At the start of 1987 another Airbus A300, G-BMNB, had been leased to Dan-Air, and appeared in the full colours of the airline. It remained with the carrier for three years, and is seen leaving London Gatwick Airport in April 1987. Towards the end of 1992 Dan-Air was taken over by British Airways, which assumed most of its scheduled operations. At this time all of Dan-Air's charter operations were suspended.

AEROMARITIME

Aeromaritime, a subsidiary of UTA, was formed as a cargo operator with its base at Paris Le Bourget. Aircraft were leased from UTA as and when required to provide passenger services. Boeing 737 Series 300 F-GFUE was in the basic colours of UTA when photographed at Manchester Airport in May 1989. Upon the merger of UTA into Air France, the fleet of Aeromaritime passed into the stock of the French national carrier.

LION AIR

Luxair and Cargolux, two airlines from Luxembourg, jointly formed Lion Air to perform passenger services on behalf of European tour operators. Services commenced in February 1988, using two Series 100 Boeing 747s originally flown by Pan Am, both configured to provide 505 economy seats. One of Lion Air's 747s, LX-GCV, is seen arriving at Manchester Airport in May 1988, while operating for Orion Air, and sister ship LX-FCV is seen in the background, having arrived at the terminal earlier that morning. At the time the airliner was operating flights out of Manchester to the island of Barbados, and carried titles to that effect.

In the summer of the following year Lion Air was still operating in the charter market, and LX-GCV had the original Orion Air fuselage titles replaced by Caribbean Airways titles. Flights were again being operated to Barbados on behalf of UK tour operators, and Manchester Airport is again the location for this May 1989 picture. Both 747s were sold during 1990, and Lion Air's charter operations were suspended.

LTS

Formed as a charter airline in 1984, to carry passengers from southern Germany to holiday destinations, LTS had its base at Munich and was a subsidiary of Düsseldorf based airline LTU. Services commenced with a fleet of two Boeing 757s obtained new from the manufacturer, further deliveries of the type following later. The first flight operated from Munich to the island of Ibiza, and it was at this Spanish island that D-AMUR, the only aircraft flown by the airline at the time, was photographed in August 1984. The LTS colours disappeared when the airline was renamed LTS Sud and the original fleet of 757s was transferred to LTE International Airways, the Spanish carrier associated with LTU.

KEY AIRLINES

Key Airlines was a Utah company based in the state's capital, Salt Lake City. Originally, general charter and contract services were flown, Cessna and Convair aircraft being used to perform these duties. Four Boeing 727s were purchased in late 1983, and these remained with the airline when its propeller-driven aircraft were disposed of at the end of 1986. Disused and parked at Las Vegas Airport in April 1985, awaiting a buyer, are Convair 440s N27KE and N28KE.

KEY AIR

The airline later moved its base to Las Vegas, introduced a new colour scheme and shortened its titles to Key Air. Boeing 727 N28KA, built in 1965 and carrying the revised livery of the airline, is seen turning on to the runway as the sun sets over the gambling city of Las Vegas in April 1985. This and nine other 727s remained with the carrier until Chapter 11 bankruptcy protection was filed for at the beginning of 1993, when the fleet was replaced by two leased McDonnell Douglas MD-83s, enabling Key Air to continue operations for a little longer. All flights were discontinued in the summer of 1993.

WORLDWIDE AIRLINES

Chicago based charter airline Worldwide Airlines began operations in the spring of 1984 with a fleet of 1969-vintage Boeing 707s originally flown by Trans World Airlines. After less than two years in business, the last of Worldwide Airlines' remaining 707s was taken out of service and the airline ceased operations. Boeing 707 N8733 is about to touch down on Miami's north runway in October 1984, bringing in a tour company's holidaymakers.

INTERESTATAL DE AVIACION

Interestatal de Aviacion, a Mexico City based airline founded in 1980, provided scheduled services within Mexico using a fleet of Convair and Vickers Viscount aircraft. Its fleet did not expand during the years of operation, and the company ceased flying when its aircraft were sold at the end of 1984. Convair 440 XA-JIR was photographed at Miami International Airport in October 1984.

SHILLELAGHS

The full name of Shillelaghs is Emerald Shillelagh Chowder and Marching Society, a Washington DC based travel club. The whole 'fleet' can be seen in this picture of Lockheed Electra N125US, photographed at Miami in October 1982. This aircraft has carried the same registration since its first flight for Northwest Orient in October 1959, having had five further owners before going to Shillelaghs in January 1972. The airline ceased operations in 1984.

NORTH EASTERN INTERNATIONAL AIRWAYS

Based on Long Island, New York State, North Eastern International Airways provided services from Islip Airport, originally with a small fleet of McDonnell Douglas DC-8s which was later supplemented by the acquisition of two ex-Pan Am Boeing 727s. The carrier's flights were mainly centred on the eastern seaboard of the USA, taking in New York, Florida and Boston. Boeing 727 N357PA, wearing the livery depicting a blue sky with white cloud formations, is seen lifting from the Fort Lauderdale runway in October 1984.

NORTHEASTERN

McDonnell Douglas DC-8 N800EV was the first aircraft to be leased by Northeastern, joining the airline in January 1982 after starting its life in the colours of United Air Lines. The Series 52 model was photographed at Islip Airport in May 1983, while it was being prepared for duty. It remained with the carrier until it was withdrawn for scrap at the beginning of 1984. Northeastern continued flying through 1984, but circumstances made it necessary for the airline to cease operations in January 1985.

MIDWAY EXPRESS

Set up by Midway Airlines in September 1984, Midway Express took over the services of failed airline Air Florida, to provide services linking its base in Chicago with points in Florida and Washington DC. An all-Boeing fleet of 737s was employed on flights in and out of these destinations, and one of the airline's aircraft is seen about to land at Miami in October 1984. By the summer of 1985 all of the Midway Express 737s had been returned to the leasing companies, their flights having been combined with those of Midway Airlines.

QUEBECAIR

The title Quebecair was adopted in 1953, and it was from Montreal that the airline provided inclusive tour and charter operations to points within Canada and the USA, and in the Caribbean and Europe. Its long-haul flights were made with McDonnell Douglas DC-8 Series 63 aircraft, and one of its fleet, C-CQBA, was photographed whilst parked and undergoing attention at Miami in October 1984, before returning to Canada. The name Quebecair disappeared in 1987, when most of the fleet was taken over by another Montreal based company, Inter-Canadian.

BRITISH MIDLAND

British Midland has been around since its formation in 1938, when it was known as Derby Aviation Ltd. The present title was adopted in 1964, and the airline is presently a successful United Kingdom carrier operating both scheduled and charter flights. This photograph, showing Vickers Viscount G-BFZL at East Midlands Airport in September 1984, depicts the British Midland livery that was in use before the current scheme, which was introduced by the airline in late 1985.

PIEDMONT AIRLINES

Piedmont Airlines was quite a large American airline and, at its close, had a fleet in excess of 150 airliners. Its base was in North Carolina, and scheduled services were operated through most of the cities in the United States. Piedmont's major hub was at Charlotte, where more than 100 flights passed daily. In October 1984, when this Boeing 727 was photographed on final approach to Miami Airport, more than 30 other examples of this popular Boeing product carried the Piedmont colours. The last flight to be operated by Piedmont occurred in August 1989, after which the airline was integrated into American airline USAir.

REPUBLIC AIRLINES

Formed in 1979 by the merger of North Central Airlines and Southern Airlines of Atlanta, Georgia, Republic Airlines had its base in Minneapolis. At one time the carrier had more than 130 McDonnell Douglas DC-9s in service, providing flights to many destinations in the United States, Canada and Mexico. This photograph of Boeing 757 N601RC at Los Angeles in April 1986 illustrates the final livery of Republic, introduced by the airline at the end of 1984. It was said at the time of its introduction that the modern red and grey design would take the airline through to the 21st century. This was not to be the case, however, as the carrier was merged and integrated into Minneapolis based Northwest Airlines soon after the picture was taken. This 757 now flies in the colours of America West Airlines.

PACIFIC WESTERN AIRLINES

Pacific Western Airlines had its base in Vancouver, Canada, and following its formation in 1945 it expanded to such an extent that it became the nation's third largest airline. While most of the scheduled services provided were centred on points within Canada, passenger and cargo charters were also flown. Boeing 737 C-GEPW came into the Pacific Western fleet brand new in November 1978, and remained with the airline until its merger with Canadian International in April 1987, with whom its still flies. Las Vegas is the location for this April 1985 picture.

BRANIFF INTERNATIONAL

After services were suspended in September 1989, Braniff did eventually get airborne once again, when operations resumed in December 1990, flying charter services only. By the summer of 1991 three Boeing 727s had been leased and scheduled services started, mostly on the eastern side of the USA. Braniff applied individual liveries to its fleet of 727s, as it had in its earlier years, adding INTERNATIONAL after the Braniff name. In fact, their scheme reminded observers of the earlier days of the airline, when 'Flying Colors' were to be seen. Here, 727 N8857E makes an early morning departure from Miami in April 1992, looking like a million dollars in its metallic gold livery.

The same service being performed by sister ship N4750, which catches the low early morning sun as it turns on to runway 12 at Miami in April 1992.

Wearing a purple and natural metal colour scheme, 727 N8855E prepares to depart Miami in June 1992. Braniff's attempts to expand suffered setbacks, competition was tough and the airline was losing passengers to other carriers. At the beginning of July 1992 all operations ceased and Braniff yet again passed into history.

NATIONAIR CANADA

Founded in 1984, Nationair Canada made its first flight in December of that year, a DC-8 service from Montreal to the island of Haiti. Operating from its Canadian base at Mirabel Airport, the airline's original fleet consisted only of McDonnell Douglas DC-8s, providing charter services within the USA and Canada, and to the Caribbean and Europe. While operating a charter flight to Canada in June 1988, DC-8 C-GXMR is pushed back from its London Gatwick gate. At that time the carrier was offering a Montreal—Brussels round-trip fare of $299, not a bad deal at 4 cents per mile.

With the introduction of the Boeing 757 to the Nationair fleet in the summer of 1990, a new colour scheme appeared. Two of the type joined the fleet, both having previously flown in the colours of Odyssey International, another Canadian charter airline that went out of business earlier the same year. The 228-seat, extended-range model 757 C-GNXI was photographed at Fort Lauderdale Airport in December 1991. Nationair continued to operate through 1992, but ran into financial difficulties at the start of 1993, ceasing all operations in March of that year.

BRITISH CALEDONIAN

British Caledonian's long-haul services were flown by the airline's fleet of McDonnell Douglas DC-10s, which joined the airline in the early months of 1977. These served on its Far East and American routes. The DC-10 G-BGAT, a Series 30 model, was photographed as it was about to depart Manchester in January 1985.

British Caledonian was created when it took over British United Airways in 1970. It provided long-haul services from the United Kingdom to West Africa and destinations within Europe, and gradually introduced cities in the USA. Charter flights were also operated. It was Britain's second scheduled airline, and British Caledonian aircraft were to be seen in various parts of the world, ranging from Hong Kong in the east to Rio de Janeiro and Texas in the west. Before the introduction of wide-bodies, long-haul services were performed by Boeing 707s, of which the airline had nine when G-BAWP was photographed at Manchester in the summer of 1974.

HELICOPTER AIRLINK

Before the construction of the M25 orbital motorway round London, it was possible to use the helicopter airlink service linking the two London airports of Heathrow and Gatwick. Introduced as a connecting service for passengers flying into one airport and out of the other, it was operated jointly by the BAA, British Airways and British Caledonian. Sikorsky S-61N G-LINK is seen at Heathrow awaiting passengers (including the author) for its next service to Gatwick in June 1980. With the opening of the M25 the helicopter services were withdrawn, as the motorway provided quick and convenient connections between the two airports.

This March 1988 photograph, taken at Manchester Airport, shows BAC One-Eleven G-AXYD in the colours of Dan-Air, operating the Manchester-Gatwick service. The twin-jet was on short lease to British Caledonian, and was returned to Dan-Air later that month. British Caledonian disappeared from the skies in early 1988, when British Airways took over the airline and its fleet.

TRANSAVIA

With the recent introduction of another new livery, Dutch airline Transavia has had two completely restyled schemes since this one, seen on Boeing 707 PH-TVA in the evening sunlight at Manchester in October 1973. Transavia colours have always included green. The airline was founded in 1965 and is based at Schiphol Airport, Amsterdam. The Boeing 707 has long since gone, and the carrier now operates a modern fleet of Boeing 737s and 757s.

EASTERN

Throughout its 63-year history Eastern had been one of the giants of commercial aviation in the USA, with routes covering the whole of the Americas. Services started in 1928. The airline was the first to put the Boeing 727 into service, the first example to appear in the Eastern livery being delivered in November 1963. The type was still flying with the carrier almost 30 years later. A Series 25 727 delivered to Eastern Airlines in December 1965, N8142N was photographed in August 1976 on final approach to Montego Bay Airport, Jamaica. During 1976 the airline applied specially prepared stickers to its fleet, and these can be seen in this picture, positioned below the cheatline. They commemorated the bicentennial year of the USA, in keeping with the 'Spirit of 76'.

Eastern was also one of the major airlines to launch the L-1011 TriStar the widebody manufactured by Lockheed, putting its first 'Whisperliner' into service in April 1972. More than twenty of the 300-plus-seat airliners were flown by the carrier, and this one, N322EA, came into service in October 1973 and was photographed arriving at Miami in December 1988.

Eastern also had the European-built Airbus A300 in its large jet fleet, the type having entered service on the eastern seaboard scheduled flights in December 1977. Airbus N225EA, delivered to the airline in October 1981, is seen turning on to runway 9L at Miami International Airport just as the sun sets on a December evening in 1990. This was to be the last Christmas for Eastern Airlines, as the company ceased all operations in January 1991 after several months of uncertainty.

In 1985 Eastern Airlines introduced services to Europe after being awarded the Miami-London route. Three long-range Series 30 McDonnell Douglas DC-10s were purchased from Italian airline Alitalia to operate these flights, and the photograph shows N392EA, after being repainted in the Eastern colours, taxying for take-off at Miami in December 1988.

SOCIÉTÉ ANTILLAISE DE TOURISME ET DE TRANSPORT

This Boeing 707, F-OGIV, is operating for the charter company Société Antillaise de Tourisme et de Transport (SATT), a French West Indies airline flying from its base at Pointe-a-Pitre, Guadeloupe, to destinations in Central and South America, bringing tourists from Europe. Two ex-Pan Am 707s were in the fleet, and the Boeing pictured in this August 1979 shot is seen at the company's base in Guadeloupe. SATT was formed in the summer of 1978, but ran into difficulties, its two aircraft being seized by the French authorities.

PAN AVIATION

Pan Aviation was the owner and operator of this executive Boeing 720. Built in 1962, N92GS came to the Miami-based charter and leasing company in the summer of 1986, and was photographed in December 1991 at its home base after being withdrawn from service. The company was formed in 1981 and had both passenger and freight aircraft in its fleet, but its operations were suspended in 1992.

TRANS AUSTRALIA

Trans Australia came into being in 1946 and retained that name until the airline became Australian Airlines in August 1986. Originally based in Melbourne, Victoria, TAA, as it was known, had a route system covering the vast continent of Australia. McDonnell Douglas DC-9 VH-TJJ, photographed at Melbourne in April 1989, still wears the 1979 TAA livery, although the first new Australian Airlines colour scheme appeared two months before the carrier's change of name, applied to a 737 rolled out of the Boeing factory.

ULTRAIR

Regular flights by Ultrair started in January 1993, operating out of its Houston, Texas, base. Five leased Boeing 727s were put into 'business class' service on flights to both the east and west coasts of America. The airline quickly built up several new routes during its first few months of operation, but found competition too fierce to continue and ceased scheduled flights after only six months in the air. Services on a much reduced scale resumed in November 1993 and, again, new services were added to the following months, but once more Ultrair found the competition too hard, suspending operations in July 1994. Being prepared for the start of initial services in December 1992, Boeing 727 N728VA is seen in the company's very attractive livery.

ORION AIRWAYS

Orion Airways was set up to fly the passengers of a United Kingdom travel company to holiday destinations in Europe, including Spain and the Greek islands. With its base at East Midlands Airport, operations commenced in the spring of 1980, using a fleet of Boeing 737s. Photographed while operating a sun-seeker charter, one of Orion's 737s, G-BHVH, is about to touch down on Ibiza's runway in August 1984.

As the travel business boomed, particularly in the peak summer months, larger aircraft were needed to transport holidaymakers. Orion's choice was the Airbus A300, and two were put into service in the summer of 1987. Each Airbus increased the carrying capacity by 328 passengers. Photographed approaching its Manchester Airport gate in October 1987, G-BMZL still wears the basic colours of its previous owner, Lufthansa.

In the United Kingdom's quiet winter periods, airlines with spare capacity keep their aircraft working by leasing them out to other carriers. Orion was no exception, and leased Boeing 737 G-BGTY to Chicago based Midway during the winter of 1987/88. Still wearing the colours of Orion, but carrying small MIDWAY titles, the twin-jet is seen about to land at Miami in December 1987. This was the last Christmas that Orion's colours were to be seen, as the travel company which originally set up the airline was acquired by the Horizon Travel Group, parent company of Britannia Airways, and the fleet came under its control.

JETAIR

Three Boeing 727s were obtained from Panama upon the start of services by Jetair, the Munich, Germany, based charter company. Flights started in the autumn of 1984, and the 727s, originally flown by Air Panama, now served new destinations in Europe and the Mediterranean. After twelve months of apparently successful operations, Jetair was declared bankrupt and the airline ceased flying in December 1985. In this picture, taken in October 1985, D-AJAA is about to depart Frankfurt/Main.

AERO VIRGIN ISLANDS

Based in St Thomas, on the US Virgin Islands, Aero Virgin Islands began operations in 1977 with a small fleet of Douglas DC-3s, performing frequent passenger services between the islands of Puerto Rico and the US Virgin Islands. About to depart Puerto Rico's San Juan airport is DC-3 N25651, built in 1940. The airline continued to use the DC-3 until it ceased operations in 1990.

CANADIAN PACIFIC AIR LINES

CP Air, or Canadian Pacific Air Lines, to give the airline its full title, was started in 1942 by the Canadian Pacific Railway and operated under this name until 1986. Its routes included domestic services within Canada and international flights to the USA, Europe and the Far East. In 1986 the airline was renamed, reverting to the original 1942 title of Canadian Pacific. At this time a new colour scheme was introduced, replacing that worn by McDonnell Douglas DC-10 C-GCPG, seen here about to depart Toronto in June 1982. The airline is currently known as Canadian/Canadien, and this Series 30 DC-10 still flies with the carrier.

AERO MEXICO

McDonnell Douglas DC-10 XA-DUH, seen at the start of its take-off roll at Miami in August 1978, illustrates Aero Mexico's natural-metal look, introduced by the airline as a cost-saving exercise. Not only can money be saved by eliminating the cost of paint, but weight is also considerably reduced, with consequent reductions in the cost of fuel used. This colour scheme was retained until the present blue and orange livery was introduced.

TAMPA AIRLINES

Tampa Airlines is still operating, providing scheduled services between its Colombian base of Medellin and Miami, although this McDonnell Douglas DC-6, HK-1276, was withdrawn from service soon after the airline's introduction of jet aircraft. The 1953-built aircraft is seen in Miami Airport's cargo area in August 1978, sporting a livery that was not to appear on the Boeing 707s which came to Tampa in 1980. This freight-only airline has now added McDonnell Douglas DC-8s to its Boeing 707 fleet.

RICH INTERNATIONAL

Rich International began operations in 1971, and was later licensed to fly all-cargo services to destinations in the USA, Canada and Mexico. Based at Miami, the airline had a small fleet of five aircraft, consisting of Douglas DC-6s and Curtiss C-46s. One of its C-46s, N74173, photographed at Miami in August 1978, shows the airline's livery at that time. Rich International is now a well-known American charter passenger airline, operating McDonnell Douglas DC-8s and Lockheed L-1011 TriStars.

FAUCETT

Faucett is Peru's leading domestic passenger airline, operating out of the country's capital, Lima. The airline has been around for quite some time, having been founded in 1928 by American Elmer J. Faucett. Its major North American destination is Miami, where McDonnell Douglas DC-8 OB-R-1257, sporting the airline's bright orange colour scheme, was photographed in October 1988. This livery is no longer used, and the DC-8s have been replaced by Lockheed L-1011 TriStars on the Miami route.

WARDAIR

Formed in 1952, Wardair commenced operations in 1953, entering the North Atlantic charter market. Services to destinations in the USA and Europe were provided from Wardair's base in Edmonton, Alberta, Canada, using a modern fleet of Boeing 747s and McDonnell Douglas DC-10s. England is a popular destination for Canadians, and the carrier's aircraft were frequently seen at Manchester Airport in the summer months. This DC-10, C-GXRB, had just left the gate at the north of England airport in June 1979, at the start of its return journey to Canada. In January 1990 Wardair was merged into Canadian Airlines.

VASP

Blue has been the basic colour used by the Brazilian airline VASP for a number of years. The carrier, owned by the state government, is based in São Paulo and currently operates an all-jet fleet of Boeing 737s, Airbus A300s and McDonnell Douglas MD-11s which carry the new livery introduced in October 1985. Boeing 737 PP-SMV, showing the scheme in use before repainting, was photographed at Rio de Janeiro in April 1983.

Also seen at Rio in April 1983 is Boeing 727 PP-SNE, finished in the darker blue and black colour scheme. At the time this picture was taken VASP had six Super 200 models in use, performing domestic flights within Brazil.

AIR 1

Air 1 started operations in 1983 with a fleet of seven Boeing 727s, providing scheduled passenger services from its St Louis, Missouri, base. It offered first-class-only travel, its 727s being configured to seat either 68 or 78 passengers, and provided complimentary wines with meals served one course at a time. The first route operated by Air 1 was to Dallas-Fort Worth, and further destinations soon followed, with the same quality of service being offered on all flights. One of the other cities flown to was New York, where N4612 is seen taxying to its gate at JFK in May 1983. Air 1 filed for bankruptcy protection under Chapter 11 in October 1984, and later ceased operations, its 727s being offered for sale.

EAST AFRICAN AIRWAYS

Douglas DC-3 5Y-AAF was one of five flown on domestic services by Kenya based East African Airways, which collapsed at the end of 1976 and was re-formed in 1977 by the government of Kenya, operating under the name Kenya Airways. The DC-3 was photographed at Nairobi in August 1975. Kenya Airways still operates today, providing scheduled domestic and international services with a modern jet fleet.

TRANSAIR

Seen at Tenerife Airport in April 1975 is Boeing 727 SE-DDA, wearing the colours of Swedish airline Transair. Founded in 1950 as a newspaper distributor, Transair was taken over by SAS in 1975 and its four 727s were put on domestic and charter operations, flying for Scanair. Services continued until Transair ceased operations in September 1981 and its aircraft were sold.

BAVARIA

One of western Germany's charter airlines was Bavaria, which had a fleet consisting entirely of BAC One-Eleven twin-jets, used on inclusive tour flights from cities in Germany to holiday destinations in Europe. Seen here at Tenerife in April 1975, D-AISY was delivered to the airline in April 1970. In March 1977 the carrier merged with Germanair to become Bavaria-Germanair.

OVERSEAS NATIONAL AIRWAYS

Overseas National Airways (ONA) was a New York Kennedy based carrier specialising in inclusive-tour and charters, mainly to Europe, using a fleet of McDonnell Douglas DC-8s, and a small complement of DC-9s to provide freight services within the USA. The DC-8s were gradually withdrawn from use towards the end of the 1970s, to be replaced by DC-10s, which continued flying for ONA until all three in their fleet were sold in October 1978. Bringing tourists from North America in May 1976, DC-8 N867F comes to stand at Las Palmas Airport.

INTERCONAIR

Licenced only to carry livestock, Irish airline Interconair obtained a Bristol Britannia and put it into service in January 1977. Previously flown by the Royal Air Force, EI-BBY was parked in the cargo area of Manchester Airport when photographed in February 1977. The airline ceased operations when the aircraft was written off in September 1977.

FLYING TIGERS

The Airfreight Airline — Flying Tigers was originally incorporated in 1945 and took the name Flying Tigers in early 1946. It was not until 1969 that the airline began world-wide cargo operations after being awarded routes across the Pacific, having previously concentrated on the USA. Flying Tigers' fleet consisted mainly of McDonnell Douglas DC-8s, later joined by Boeing 747F freighters. Jumbo N808FT is seen at JFK Airport, New York, in May 1981, proceeding for take-off. The DC-8s were later taken out of service and the 747s were supplemented by 727s, which performed internal freight services. The name Flying Tigers started to disappear from the skies in August 1989, when the business was taken over by Memphis based freight airline Federal Express.

BAJA CORTEZ AIRLINES

Based at Los Angeles International Airport, Baja Cortez Airlines commenced services with a single Riley D.H. 114 Heron, N690BC, seen here at LAX in August 1977. This aircraft was later taken out of the fleet and replaced by five Piper Navajo Chieftains which provided scheduled passenger services to points on Mexico's Baja California peninsula. Flights continued until the carrier ceased operations in April 1980.

ALLEGHENY COMMUTER

De Havilland Twin Otter N103AC, awaiting passengers at La Guardia Airport, New York, in May 1983, carries the titles Allegheny Commuter, the operating name for Atlantic City based Southern Jersey Airways. Services started in June 1970, linking Atlantic City with Washington DC, Trenton and New York. Allegheny Commuter services were also operated by six other regional carriers, providing a feeder connection for USAir at major airports. Allegheny aircraft are now operated in USAir colours as USAir Express.

COMMAND AIRWAYS

Command Airways was another of the east coast of America's commuter airlines providing scheduled passenger services from upstate New York to the airports of La Guardia and JFK. One of the carrier's Shorts 330s N939MA, was photographed while visiting Boston in May 1984. In 1991 Command Airways merged into Flagship Airlines and operated as American Eagle, the commuter airline providing connections for American Airlines.

BRITISH ISLAND AIRWAYS

Originally formed in 1971, British Island Airways was merged into the Air UK group and then operated independently out of Gatwick Airport, London, providing charter flights for the UK travel organisations. BAC One-Elevens formed the majority of the carrier's fleet, remaining with the airline until, owing to a declining charter market in the UK, it found it necessary to start disposing of them. More assets were sold, and the airline finally ceased all operations in February 1990. G-AYWB is seen about to depart Palma, destined for London Gatwick, while operating holiday charter flights in May 1986.

TRANS INTERNATIONAL AIRLINES

A subsidiary of the Trans America Corporation, Trans International Airlines had a base in Orlando, Florida, from where charter services were provided using their sole passenger-carrying aircraft, a McDonnell Douglas DC-8, and two Lockheed L-188s for freight services. When passenger flights ceased, the company continued operating the freighters until all flights were finally discontinued in 1990. The DC-8 N4868T was on lease to the airline from the Trans America Corporation when it was photographed operating a flight into Las Vegas in April 1986.

SKY SAFARI

Sky Safari Boeing 707 N196CA, photographed at St Petersburg airport, Florida, in May 1980, was bought by the Charlotte Aircraft Corporation in July 1978 and remained with the company until August 1981, when it passed into the ownership of Hispaniola Airways. It was finally broken up after being damaged in a landing accident at Miami in December 1981.

(Overleaf)

ASPEN AIRWAYS

Two of Aspen Airways' three colour scheme variations are shown in this picture, taken at Denver in August 1985. British Aerospace BAe 146 N461 is taxying to its stand after completing one of Aspen's Colorado services, which took the carrier's colours to most major towns and cities in the State. During the winter months, one of the carrier's popular services transported skiers to the well-known resort of Aspen. Aspen Airways began operations in 1962, and at one time had a fleet consisting entirely of Convair 580s. In 1989 the Aspen titles disappeared from its aircraft when the fleet then in operation started operations on behalf of United Airlines, wearing the colours of that carrier and displaying United Express titles.

TOTAL AIR

US charter airline Total Air was formed in 1984 and had its base at Los Angeles. The fleet consisted entirely of Lockheed L-1011 TriStars, flown mainly to North American destinations but also to European cities. The name Total Air survived only a few months, as the airline was renamed Air America. Still wearing Total Air titles, one of the L-1011s in the fleet is seen on approach to London Gatwick Airport on one of the United Kingdom's brilliant summer Saturdays in July 1987. Air America filed for liquidation in March 1990 and suspended all operations, its fleet of TriStars going into store.

BRITISH AIR FERRIES

Formed in 1963, British United Air Ferries Ltd operated passenger and vehicle services across the English Channel from its Southend-on-Sea base. The name was shortened to British Air Ferries (BAF) in 1967. Vickers Viscount G-BLNB was photographed at East Midlands Airport in August 1986. The airline continued to operate propeller-driven aircraft, and later introduced jet airliners into the fleet with the acquisition of BAC One-Elevens in 1990. In April 1993 the airline adopted a new colour scheme upon changing its name to British World Airlines.

SCANAIR

Stockholm, Sweden, based charter airline Scanair was formed in 1965 by a consortium, and was owned by three Scandinavian airlines. It performed inclusive tours to holiday destinations in the Mediterranean and the Canary Islands. Aircraft were leased from other Scandinavian carriers and operated in the full colour scheme of Scanair, shown in this picture of Airbus A300 SE-DFK about to touch down on the Balearic island of Mallorca in May 1986. The French-built airliner carries the name *Sven Viking*, given to it by SAS, to whom the aircraft belongs. Scanair leased the 291-seat twin-jet for a little over four years before replacing its entire fleet with McDonnell Douglas DC-10s. In January 1994 Scanair merged with Conair, another Scandinavian carrier.

GUERNSEY AIRLINES

The island of Guernsey, in the Channel Islands, had its own airline, Guernsey Airlines, started in 1977 to provide scheduled services from the island to the UK mainland. There was never a large fleet of aircraft, and the carrier relied on Shorts 330s and 360s and Vickers Viscounts to operate its services. In October 1989 the airline merged into Air Europe Express, and its aircraft were repainted in the livery of the new owners. Photographed at East Midlands Airport in August 1988, Vickers Viscount G-AOYG awaits passengers for its flight to Guernsey.

HISPANIA

Spanish carrier Hispania chose the French-built Sud-Aviation Caravelle with which to start operations in 1983, and put four examples in use providing charter flights out of the island of Palma. The Caravelles were all eventually replaced by other equipment, and by 1987 Hispania's entire fleet consisted of Boeing 737s. New aircraft were added with the introduction of Boeing 757s in February 1989, but these remained in the carrier's colours only a matter of months, as Hispania declared bankruptcy in the summer of that year and its aircraft were impounded at various European airports. Caravelle EC-DCN was photographed at Palma Airport in May 1986, during the airline's earlier days.

WORLDWAYS

Canadian charter airline Worldways started long-haul flights in 1981, with a small fleet of three Boeing 707s. Before this the company was a domestic Canadian carrier which expanded considerably after its 1974 formation. The three Boeings, configured to provide 195 economy seats, remained in service until they were replaced by 245-seat McDonnell Douglas DC-8s and 362-seat Lockheed TriStars, which continued operating the airline's charter services to European and US destinations. Seen parked at Toronto between duties in May 1982 is Boeing 707 C-GFLG, which remained in service with Worldways until 1985, its last operational 707. The airline ceased all operations in October 1990.

AIR PANAMA

Air Panama started operations in 1969 and became the designated flag carrier of the country. Flights from the airline's Panama base linked the capital of the country with points in North and South America. During the 1980s Boeing 727s performed these duties and, apart from the occasional lease of a McDonnell Douglas DC-10, the Boeings remained exclusively with the airline. Boeing 727 HP619 displays the Air Panama colours of 1979, and is seen about to turn on to runway 9R at Miami in August of that year.

Another of Air Panama's 727s joined the airline in October 1986, on lease from Alia Royal Jordanian Airlines. Still bearing the Jordanian registration JY-ADR, the tri-jet was photographed landing at Miami in October 1987, bringing in passengers on the carrier's regular flight from Panama. The livery is that of Alia, with the Air Panama titles and logo added. Air Panama ceased operations in January 1990, after its Miami route was taken over by Panamanian airline Copa.

AIR MIDWEST

Photographed at Denver in August 1986, this Air Midwest 17-seat Swearingen Metro II, N241AM, was operating a commuter service in the mid-western states of the USA. From its base in Kansas, the carrier's fleet of Metros flew to cities in Colorado, Oklahoma, New Mexico and Texas. The airline later concentrated on providing connecting flights on behalf of Trans World Airlines and Eastern Airlines, and it now performs these flights for US Air Express in the full colours of that airline.

TRANS-COLORADO AIRLINES

Trans-Colorado Airlines was a commuter airline providing services from Denver, the State capital, to various destinations within the State. Founded in 1979, the airline relied on its fleet of Swearingen Metros to operate its services. In 1987 the aircraft started flying on behalf of the American airline Continental, with Continental Express titles in the full colours of that airline. Metro III N31088 was photographed at Albuquerque, New Mexico, in August 1985, wearing the Trans-Colorado livery.

NORTHWEST ORIENT

Northwest Orient is one of the United States' oldest and major airlines, having been founded in 1926 as an airmail carrier operating from Detroit and its Minneapolis/St Paul base. The airline was known simply as Northwest until 1934, when a slight name change to Northwest Orient was made. This title remained until the word 'Orient' was removed in 1986, upon the merger of Republic Airlines into the company, when the carrier reverted to its original name. All aircraft in the fleet bore a similar colour scheme, although pure Boeing 747 freighters had a highly polished natural-metal finish with red Northwest Orient Cargo titles. Photographed at Manchester Airport in June 1981, soon after arriving from Amsterdam, N618US illustrates this scheme.

Northwest Orient had a number of long-range McDonnell Douglas DC-10 Series 40s in its fleet, and N160US, photographed in October 1982, is seen a few seconds before touchdown at Miami Airport. Both of the aircraft depicted are still flying for Northwest and carry the company's latest livery.

RANSOME AIRLINES

Pennsylvania based Ransome Airlines operated scheduled commuter services to various cities in the eastern United States including Washington DC, New York and Boston. Flights commenced in 1967 and continued in the colours of Ransome until 1986, when all aircraft began operating on behalf of Pan Am Express, in Pan American colours. De Havilland Canada Dash 7 N174RA was photographed proceeding to its gate at New York's Kennedy Airport while operating a Ransome commuter service in May 1983.

PHILIPPINE AIRLINES

Philippines government-owned carrier Philippine Airlines has a worldwide routeing structure operated by a modern fleet of Boeing and Airbus jetliners, together with Fokker 50s used for short internal flights. Its present colour scheme was unveiled in 1986, and features a sunburst tail design and the word 'Philippines' on the fuselage. The airline's previous livery is shown on Airbus A300 RP-C3002, being pushed back from its gate at Hong Kong Airport in April 1981. At the time the airline was introducing scheduled flights to the west coast of America, hence the added logos 'Hurrah for Hollywood' and 'The Love Bug'.

MID PACIFIC AIR

Mid Pacific Air was an Hawaiian airline which began operating in 1981 with a small fleet of the Japanese-built NAMC YS-11s, which were used on scheduled inter-island services. The airline also operated routes on the US mainland, linking cities within California and Nevada. After a spell flying commuter services for Continental Airlines as Continental Express, the airline ceased operations in 1988, later resuming services as a US mainland charter and cargo airline with the same YS-11 fleet. The airline was renamed Mid Pacific Cargo and operated from a base in Lafayette, Indianapolis. Mid Pacific Air NAMC YS-11 N115MP was photographed at Honolulu in April 1982.

AIRLIFT INTERNATIONAL

Airlift International started life as a cargo airline operating out of its Miami base following its formation in 1945. In its later years of business, freight services to points in Central and South America were undertaken, using the carrier's fleet of McDonnell Douglas DC-8s. During the following years its DC-8s were replaced by Fairchild F-27s, which provided passenger services. An attempt to auction the assets of the airline failed in 1992. Operating a freight service for Airlift, DC-8 N717UA was photographed as it was about to depart Miami in August 1979.

BRITANNIA AIRWAYS

Britannia Airways is the UK's largest charter airline, operating an all-jet fleet of Boeing airliners on both charter flights and inclusive tours from all parts of Britain to many overseas destinations. Boeing's famous 737s at one time made up the entire Britannia fleet, but they have now been superseded by the manufacturer's larger aircraft, including the 757 and 767, of which the airline now has nearly 30 examples. Part of this fleet includes the extended-range 767, which is used on Britannia's scheduled operations from the UK to Australia. The present livery of the airline, unveiled in 1983, replaced that in use during the 1970s, which is seen on Boeing 737 G-AZNZ, photographed at Malaga in March 1978.

In the summer of 1981 Britannia found it necessary to lease aircraft to enable schedules to be maintained. Boeing 737s were added, and TF-VLK appeared wearing a revised Britannia livery. Sub-leased to the airline by the Icelandic airline Eagle Air, the Boeing flew for Britannia for six months before returning to Iceland.

AIR SUNSHINE

Based in Key West, on the southern tip of Florida, Air Sunshine operated daily scheduled services to destinations within the sunshine state. The title was adopted in 1972, and for a further five years the Air Sunshine fleet consisted entirely of DC-3s. The airline's services continued until the end of 1978, when it was acquired by Air Florida. Still in the livery of Air Sunshine, Convair 440 N477KW was captured by the camera at Miami in August 1979.

UNIFLY EXPRESS

Founded in 1978, Unifly Express was an Italian airline based at Rome's Ciampino Airport. Services started using a small fleet of propeller-driven aircraft. This was added to in 1988, when the airline took delivery of McDonnell Douglas DC-9 freighters and MD-82/83 passenger carrying jets. Leased to the airline by an Irish leasing company, MD-82 EI-BTX flew in the Unifly colours until early 1990. This photograph shows it shortly after landing at Frankfurt/Main Airport in May 1988. In May 1990 the airline ceased operations.

AIR HAITI

Air Haiti began operating cargo flights from the Haitian capital, Port-Au-Prince, to Miami in 1969 with Douglas DC-6 and Curtiss aircraft, then in April 1980 the airline leased a Boeing 707, enabling it to provide a better service on the Miami route. The 707, N15711, continued to operate until January 1982, and was photographed in October 1980. Bearing the full Air Haiti titles, it is about to land at Miami. The 707 was returned to the leasing company in January 1982, and Air Haiti continued operating with Curtiss aircraft until it ceased operations in 1983.

SKYBUS

Based in Denver, Skybus was previously known as Frontier Horizon Inc and then as Frontier Discovery Airlines. It was founded in 1984, and had a fleet of Boeing 727s. As well as operating its own flights, the airline performed scheduled services on behalf of World Airlines. All operations ceased at the end of 1986, when the company went into the leasing business. Boeing 727 N545PS is seen about to take off from Las Vegas airport in April 1986, while on a charter flight.

HENSON AIRLINES

Henson Airlines was a scheduled passenger airline providing services in the US states of Maryland, Pennsylvania and New York, using a small fleet of propeller-driven aircraft. Founded in the State of Maryland in 1931, the airline inaugurated Allegheny Commuter flights in 1967, and these continued until 1983, when the airline was sold to Piedmont. Services then continued under 'Piedmont Regional' titles, and De Havilland Dash 8 N909HA was photographed in December 1988 while operating in this scheme, flying in to Miami. When Piedmont Airlines was taken over by USAir in the summer of 1989, the Henson Airlines fleet, previously seen in Piedmont Regional titles, soon reappeared in the colours of USAir.

AMBER AIRWAYS

Amber Airways Ltd commenced charter services in April 1988 with two aircraft, one of which, a leased Boeing 737, carried the Amberair name and colours. Based at Cardiff, the company operated for only a few months before being acquired by and integrated into another UK charter airline, Paramount Airways, in October 1988. This airline suffered financial problems and, unable to find sufficient business to keep aircraft in the air during the winter of 1989, folded at the beginning of 1990. Boeing 737 G-BKMS, operating in Amberair titles, is seen approaching one of the gates at Manchester Airport on a hazy June morning in 1988.

AMERICAN INTER-ISLAND

American Inter-Island was a subsidiary of American Airlines, operating scheduled services on behalf of the parent company. From its St Croix base in the Virgin Islands, passenger and cargo flights were provided to St Thomas and San Juan, using Convair 440s. Two of the carrier's fleet of five are seen being prepared for flight at St Croix Airport in August 1980. In 1982 the airline was purchased by California based Air Resorts Airlines, and further aircraft were added to the fleet of the new company. This airline is still operating, its fleet consisting entirely of Convair aircraft.

AIRWAYS CYMRU

Welsh airline Airways Cymru was based in Cardiff. Founded in 1984 as Airways International Cymru Ltd, the airline provided inclusive-tour flights to holiday destinations from Cardiff and Manchester with BAC One-Elevens and, later, Boeing 737s. These services continued for only four years, as the company ceased operations in January 1988. BAC One-Eleven G-YMRU was returning to the United Kingdom on a holiday flight when photographed as it was about to start its take-off roll at Ibiza Airport in August 1987.

AIR PUERTO RICO AIRLINES

McDonnell Douglas DC-9 Series 14 N931EA was originally delivered to Delta Air Lines in December 1965, and retained N931EA in February 1984 when it was bought by the American company Emerald Airlines. While owned by Emerald it was leased to Air Puerto Rico Airlines for three months, and it was during this period that it was photographed in December 1986, about to depart Miami.

SUN AIRE LINES

Sun Aire Lines began operations in December 1968, providing connections from its Palm Springs base to various other Californian cities. Seen at Los Angeles International Airport in August 1981, N64SA was one of a fleet of Swearingen Metros operated by the airline on feeder services. Sun Aire continued operating until 1985, when it merged into Utah based Skywest Airlines, providing scheduled feeder services on behalf of Western Airlines as Western Express.

DELTA AIR

Saab 340 D-CDIC was one of four of the type operated by Delta Air, based in Friedrichshafen, Germany. Operations began in 1978, when scheduled passenger services were initiated between the main German cities and points in Switzerland. Frankfurt/Main is the location of this May 1988 photograph. The title Delta Air disappeared when the airline was renamed Deutsche BA in June 1992.

AIR ATLANTIS

Air Atlantis was a Portuguese airline formed in 1985 to take over and increase the charter services of the country's flag carrying airline, TAP Air Portugal. Boeing 707s, 727s and 737s were leased from TAP to operate these flights from Faro Airport to various European destinations. Boeing 737 CS-TET is seen approaching the gate at Düsseldorf Airport whilst operating an inclusive tour between its home country and Germany in May 1988. Following a decision by the shareholders of TAP, Air Atlantis suspended operations in February 1993.

WESTERN AIRLINES

North American airline Western was a long-established carrier, originally being formed as far back as 1925. The airline's main base was Los Angeles, from where services to most North American cities were operated, together with flights to Mexico and Honolulu. McDonnell Douglas DC-10 N907WA is seen arriving at its allocated gate at Honolulu after flying from the North American mainland on a spring day in 1982. Western Airlines merged with Delta Air Lines in the early part of 1987, and its aircraft were repainted in the scheme of its new owners.

EMERY AND PUROLATOR

Emery and Purolator were two different companies involved in the air freight business. Emery Worldwide is still in business, and is one of the leading freight and courier companies, operating worldwide services with a comprehensive fleet of Boeing and McDonnell Douglas aircraft. Purolator Courier, as it was known, was a smaller company which provided a similar service before merging with Emery in September 1987. For a short time, Boeing 727 N418EX appeared with dual titles, as seen in the picture taken at Washington Dulles Airport in October 1988.

WORLD AIRWAYS

Charter operator World Airways commenced services in 1949 and became a leading non-scheduled airline, providing passenger and cargo services and flights under contract to the US military. Its current fleet consists entirely of McDonnell Douglas DC-10s and MD-11s, the passenger carrying versions all configured to economy-class seating. Illustrated is a McDonnell Douglas DC-8 at Manchester in March 1975, after push-back. It bears the livery in use at that time, which was replaced in 1986 by an all-white fuselage.

DOMINICANA

Dominicana was a Dominican Republic airline based in the country's capital, Santo Domingo. The company operated passenger and cargo services mainly within the Caribbean, but services to the American mainland were also provided. The Dominicana livery features red and blue cheatlines over a white fuselage, but this illustration of Boeing 727 HI-630CA is an exception as this airliner carries the Dominicana titles applied to an aircraft previously flown by and in the colours of the American airline Braniff. The 727 flew for Dominicana for a short period after the demise of the American airline, and is seen about to touch down at Miami in December 1992.

TAINO AIRWAYS

A more recent airline to start services from the Dominican Republic was Taino Airways, which operated charter flights as required. In early 1993 the airline leased a McDonnell Douglas DC-10 from African airline Scibe-Airlift Zaire, and put it into service on charter flights from the Dominican capital Santo Domingo to Frankfurt/Main. It was at this German city that the French registered tri-jet, F-GHOI, was photographed as it left for the Caribbean on a Saturday afternoon in August 1993. Taino Airways still operates, though it does not have aircraft, services being provided using leased equipment.

JET AIRE AIRLINES

Jet Aire was a commuter airline operating in and around New Mexico. Services started in October 1984 with a fleet of two Handley Page HP.137 Jetstream 1s, and N114CP, one of its Jetstreams, was photographed at Albuquerque in August 1985, about to depart. During the short time that this airline was flying, it never had more than two aircraft in its fleet. When Jet Aire ceased operations in January 1986, the 'fleet' had been reduced to just one aircraft.

INTER EUROPEAN AIRWAYS

Inter European Airways was a subsidiary of Aspro Holidays Ltd, a United Kingdom inclusive tour and holiday company. Operations began in April 1987, when leased Boeing 737-200s were put into service flying to the sunspots of Spain, Italy and Greece. When the Series 300 737s ordered earlier by the airline were delivered, the Series 200 Boeings were returned to the leasing companies. Further aircraft were added to the fleet in 1990, with the delivery of Boeing 757s. Boeing 737 G-IEAA, a Series 300 model, was photographed at Manchester in May 1993, while being pushed back at the start of its journey to a Mediterranean destination. Parked in the background awaiting its next flight is another of the company's Series 300s. The airline continued operations until September 1993, when it was taken over by Manchester based company Airtours International.

AIR WISCONSIN

Air Wisconsin operated scheduled passenger and freight services from its Wisconsin base to destinations in the mid-states of North America. Operations commenced in 1965, and the airline built up a fleet of both propeller-driven and jet aircraft. Chicago was one of the destinations served by the carrier, and it is at the windy city's O'Hare Airport that De Havilland Canada Dash 7 N708ZW was photographed in April 1985, about to depart on a commuter flight as the last rays of the sun caught the terminal building. In 1986 the airline became a scheduled feeder for United Airlines, and operated as United Express in full colours and titles.